A PICTURE BOOK OF JAMES WEBB WONDERS

A Kid's Guide to JWST's Space Discoveries

John Martinez

TABLE OF CONTENTS

INTRODUCTION

Space travelers, how are you? Think of the James Webb Space Telescope (JWST) as a really cool space detective. In our space journey, this telescope is like a superhero because it helps us find out about the universe's amazing secrets and wonders.

Think of a group of smart and interested scientists who are working very hard to make this cool telescope. They make it tough enough to handle the tasks of space by using very accurate tools and high-tech stuff. It's like having to make a superhero suit for space!

We are now ready for the big launch! The JWST quickly moves away from Earth and spreads its golden wings into space, a million miles away. Its job is to show us how beautiful the universe is.

We want you to come on a great adventure with us in our book, "Awesome Discoveries from the James Webb Space Telescope." Let's follow the JWST's story as it answers space riddles that have puzzled experts for a long time.

We'll share the latest and best findings, explaining them in a super cool way and showing you awesome pictures. The JWST takes pictures of things in space we've never seen before, like galaxies dancing far away and the birthplaces of stars hiding in clouds of gas and dust.

Together, let's discover new places where the sky looks absolutely amazing. The JWST catches these stunning pictures, and our book is like a magical door to the amazing beauty of space.

So, put on your space suits and get ready for an amazing trip! The JWST is all set, with its tools and lens ready to discover the amazing secrets of space. Turn the pages, and be ready to be surprised by the awesome gifts waiting for you. Welcome to the latest age of space travel, where we discover the secrets of the sky one jaw-dropping picture at a time!

CHAPTER ONE

WELCOME TO SPACE: THE GREAT COSMIC PLAYGROUND

Hey future astronauts! Have you ever looked up at the dark sky and thought what's out there beyond the stars? That's space! It's like the largest playground ever, but instead of swings and slides, there are planets, stars, and planets.

It's so huge and strange, and scientists are like cosmic spies trying to uncover its secrets.

Imagine being in a rocket ship, speeding past worlds and whirling galaxies. Space is where ideas become as big as the world, and our desire to explore it is like a never-ending journey.

Meet James Webb: The Space Explorer Hero

Now, let's talk about a super cool person named James Webb. He's not a character from a comic book, but he's just as awesome! James Webb was a really smart and interesting guy who loved studying space. He believed in the magic of the stars and wanted to understand everything about them.

So, scientists named a space camera after him because he pushed them to dream big and explore the unknown. It's like having a magic telescope named after a space character! The James Webb Space Telescope, or JWST for short, is like a friend helping us find the amazing secrets hidden in the far reaches of space.

What's JWST? A Cosmic Super Friend!

JWST is a special name for our cosmic buddy – the James Webb Space Telescope! Let's break it down:

J: For James, the space scientist guy.

W: For Webb, his last name, like a cool space fighter.

S: For Space, because that's where our camera friend loves to roam.

T: For Telescope, the magical space tool that helps us see amazing things in the sky.

Now, whenever you see "JWST," just remember it's our space buddy's nickname – James Webb Space Telescope!

So, put on your space gear and get ready for an amazing trip with JWST, our space fighter camera!

CHAPTER TWO

BUILDING THE TELESCOPE

Teamwork in Space

Once upon a time, a group of bright minds came together like a cosmic superhero team to build something extraordinary—the James Webb Space Telescope (JWST). Imagine scientists and engineers working side by side, just like space buddies on a mission. They were curious and excited, with a shared goal: to build a telescope that could explore the secrets of the world.

These space experts had to be like puzzle masters, figuring out how to make every piece fit correctly. Each person had their own special skill, whether it was planning, making, or testing. It was teamwork at its best, with everyone sharing ideas to make the JWST the best space detective ever.

The U.S. Department of Defense makes sure the James Webb Space Telescope gets important stuff it needs!
Creator: Courtesy Photo, Credit: U.S. Civilian,
Copyright: Public Domain

Picture this: scientists thinking up cool plans, engineers making exact pieces, and everyone working together like a space family. They faced obstacles, fixed problems, and never gave up. It was like building a space rocket, but even cooler—a camera that would help us see things in space like never before!

Amazing Technology

Now, let's talk about the JWST's superpowers—its amazing technology! This telescope is like a space wizard with the coolest toys. First off, it has giant screens that act like cosmic nets, grabbing the tiniest bits of light from far, far away. These mirrors are so big and special that they can see things our eyes can't.

But that's not all! The JWST is packed with cutting-edge tools and sensors. Imagine having a space camera that can record pictures in super-duper detail. The telescope's tools can study light, helping scientists understand the secrets hidden in the colors of the sky.

It's not just about getting shots; the JWST can also tell us what things are made of in space. It's like having a space traveler with a great sense of smell, sensing the cosmic ingredients of stars, planets, and other space wonders.

As the scientists and builders put together this amazing camera, they made sure it could withstand the difficulties of space. It's tough, like a space fighter suit, ready to face the cold, the heat, and everything else space throws at it.

So, in Chapter 2, we're digging into the teamwork that made the JWST possible and exploring the mind-blowing technology that turns it into a space fighter ready to find the universe's biggest secrets!

CHAPTER THREE

LAUNCHING INTO SPACE

Preparing for Launch

The James Webb Space Telescope (JWST) is like a brave space warrior getting ready for an exciting journey! Before it can zoom off into space, a team of super-smart scientists and engineers has to make sure everything is just right.

Space Suit Check: Imagine the JWST wearing a special space suit! Scientists have to make sure it can handle the tough conditions in space, like high cold and hot temperatures.

Packing for the Journey: Just like you pack your bags for a trip, the JWST has to be packed carefully. Scientists make sure it has all the tools and instruments it needs to explore the universe.

Safety Inspections: Before any space journey, there are safety checks. Scientists try and double-check every part of the JWST to make sure it's super strong and can handle the challenges of space travel.

Final Pep Talk: Just before launch, everyone meets for a pep talk. They encourage the JWST, telling it to be brave and do its best in discovering the wonders of space.

Countdown to Blastoff

Get ready for the big moment! Launching the JWST into space is like sending a superhero off on a journey. Here's how it all happens:

T-Minus Countdown: Picture a giant clock ticking down to launch. The countdown starts, and everyone is super excited. It's like waiting for a rocket to blast off into a fantastic space journey.

Rocket Power: The JWST doesn't have wings like a bird; instead, it rides on a powerful rocket. When the countdown hits zero, the rocket engines roar to life, sending the JWST higher and higher.

avoid from Earth's Grasp:** Earth has a strong hug called gravity, but the JWST needs to avoid it. The rocket helps it break free from Earth's grasp, and the telescope starts its trip to explore the vastness of space.

Flying to the Cosmos: The rocket takes the JWST on a million-mile trip. It's like an exciting ride through space, and the camera is on its way to spread its golden wings and start its cosmic mission.

So, as the countdown hits zero and the rocket engines roar, the JWST begins its grand trip, ready to discover the secrets of the universe. Buckle up and get ready for the ride of a lifetime!

CHAPTER FOUR

DISCOVERING THE NATURAL WORLD

Finding out the Secrets

Get ready for some mind-blowing secrets of the world! The James Webb Space Telescope (JWST) is like a cosmic detective, and it's on a mission to reveal secrets that have puzzled science for ages. Let's sneak a peek at some of the exciting findings the JWST is aiming to make:

1. Searching for Alien Atmospheres: The JWST is like a space detective with a special tool to study the atmospheres of worlds outside our solar system. Imagine finding hints about strange worlds and what they're made of!

2. Time Traveling with Light: This awesome telescope can see really far back in time by catching light from super faraway galaxies. It's like looking at cosmic time capsules and learning what the world was like when dinosaurs walked the Earth.

3. Chasing Ghostly Stars: The JWST will track down the lightest, most mysterious stars in the sky. These stars are like cosmic ghosts, and the telescope will show their secret stories.

Amazing Images

Now, prepare to be stunned by the amazing pictures the JWST records. It's like having a front-row seat to the best cosmic show ever! The telescope's powerful eyes will unveil stunning scenes:

1. Galaxies Doing the Tango: Watch as distant galaxies spin and dance in fascinating cosmic dancing. The JWST will record these cosmic dances, showing us the beauty of galaxies far, far away.

Source: X-ray (Spitzer): NASA/CXC/SAO; IR (Webb): NASA/ESA/CSA/STScI; IR (Spitzer): NASA/JPL-Caltech; Stephan's Quintet: Webb image data already merged with the Spitzer telescope

Galaxy in Cartwheels. Webb, Chandra, composite, from top to bottom. Credit: IR (Webb): NASA/ESA/CSA/STScI; X-ray: NASA/JPL-Caltech; CXC/SAO: NASA

NASA

Look at this cool picture from the James Webb Space Telescope! Some smart scientists at UCLA did studies to learn new things using this amazing telescope.

Look at this cool picture from the James Webb Space Telescope! It's the center of M74, also called the Phantom Galaxy. Credit:
ESA/Webb, NASA & CSA, J. Lee and the PHANGS-JWST Team.

Take note

M74: Messier 74 (M74) is a spiral galaxy located in the constellation Pisces. It's a beautiful galaxy with spiral arms, where stars twirl like a cosmic dance. Scientists love studying M74 to learn more about the mysteries of the universe!

The Phantom Galaxy: This is another name for Messier 74 (M74), which is a spiral galaxy. It got its nickname because of its faint appearance, making it look a bit mysterious and ghostly. Scientists use telescopes like the James Webb Space Telescope to take cool pictures of this galaxy and learn more about its secrets!

2. Nebulae: Clouds of Cosmic Colors: Get ready for a feast of colors! Nebulae are like art galleries in space, and the JWST will paint detailed pictures of these clouds made of gas and dust, where new stars are born.

The captured image of nebula L1527, taken by Webb's Near-Infrared Camera (NIRCam), includes reference compass arrows, a scale bar, and a color key.

Look at this amazing picture! It shows a place called NGC 3324 in the Carina Nebula, where stars are born. The James Webb Space Telescope took this special picture in a way that lets us see parts we couldn't before. It's like peeking into hidden star secrets! NASA, ESA, CSA, and STScI

Look at the Pillars of Creation through NASA's James Webb Space Telescope! This cool picture was taken using special light. **NASA, ESA, CSA, STScI; J.**

3. Planets in the Spotlight: The telescope will focus on planets in our own solar system, highlighting their unique traits. Imagine seeing the detailed features of Jupiter's giant storms or Saturn's stunning rings up close!

The James Webb telescope took a cool picture of Fomalhaut, a bright, young star 25 light-years away! Guess what? There might be hidden planets hanging out near it! (NASA/ESA/CSA/A. Pagan/A. Gáspár via CNN Newsource

So, tighten your space harnesses as we start on this amazing trip with the JWST. Each page turn will show a new cosmic surprise, bringing the wonders of the universe right to your hands. Get ready to be surprised by the amazing pictures and finds waiting for us in the vastness of space!

Webb's special camera took a cool picture of Jupiter using red, yellow-green, and cyan filters. The colors show Jupiter spinning around!

Credit: NASA, ESA, CSA, Jupiter ERS Team; image processing by Ricardo Hueso (UPV/EHU) and Judy Schmidt.

CHAPTER FIVE

HOW JWST WORKS

Light and Telescopes

Hey space lovers! Let's enter into the beauty of light and binoculars. Imagine light as a cosmic servant bringing notes from distant stars and galaxies. Telescopes are like super eyes that help us notice those signs!

1. Light Messages

In space, there's a lot going, but we can't see it with our normal eyes. That's where binoculars come in. They gather and focus light, making it into pictures we can see. It's like using a huge space magnifying glass!

2. Cosmic Light Show

Stars, planets, and galaxies shine bright because they give off light. Telescopes grab that light and bring it closer to us. It's a bit like turning up the light on your space TV!

3. Seeing Far and Clear

Telescopes help us see things that are super far away. They gather more light than our eyes, making faraway things sharper and more detailed. It's like having superhero vision for exploring the stars.

JWST's Superpowers

Now, let's talk about the James Webb Space Telescope's (JWST) super cool skills that make it an amazing space detective!

1. Giant Mirror

JWST has a huge screen, much bigger than any other space camera. This large mirror is like JWST's superhero eye, catching tons of light to see weak and faraway things. It's like having a cosmic torch that can reach quite far!

2. Infrared Vision

JWST doesn't just see normal light; it can also see infrared light. Infrared is like a secret code that some things in space use to interact. With this special power, JWST can reveal secret riddles, like finding young stars in dusty clouds.

3. Sunshield Super Suit

Imagine a fighter with an awesome shield that protects it from the sun's rays. Well, JWST has a sunshield, like a huge umbrella, that keeps it cool and protects it from the sun's heat. This helps the camera to focus on its task without getting too hot.

4. Moving Parts Dance

JWST has cool, moving parts that can dance into different poses. This dance helps it view different parts of the sky. It's like having a character who can do gymnastics to get the best view of the action!

In a word, JWST is a space fighter with a big eye, special vision skills, a sunshield, and some amazing dance moves. With these skills, it's all set to explore the world and bring us mind-blowing pictures of the cosmic wonders beyond!

CHAPTER SIX

INTERESTING FACTS AND ACTIVITIES

Did You Know?

1. Super Mirror Magic: The James Webb Space Telescope has unusual mirrors that are like superheroes! They're extremely huge and can collect more light than any other space camera. It's like having a big space screen to see afar stars as well as planets.

2. Freezing in Space: Imagine a camera that gets really, really cold! The JWST's sensors need to be super cool to work properly. It's like having a space telescope with its very own freezer!

3. Time Traveler's Eye: The JWST can see back in time! It can catch the light that's been moving for billions of years. It's like having a time machine, but with pictures instead of buttons.

4. Alien Sunscreen: Space is full of powerful sunlight, but the JWST has a special shield, like a giant space cover, to protect its delicate sensors from the Sun's rays. It's like giving the camera its own cool space sunscreen!

Become a Space Explorer

1. Build Your JWST: Get creative and make a small James Webb Space Telescope using items like cardboard, foil, and plastic. Imagine how experts built the real one and create your own space fighter!

2. Spy on Space Objects: Pretend you're a space spy! Use a small telescope or glasses to view the Moon, stars, or even planets. Draw what you see in a space explorer's notebook.

3. Cosmic Colors: Learn about the colors of stars! Use art tools to draw different stars with unique colors. Did you know stars come in red, blue, and even gold?

4. Alien environment Experiment: Create your own "alien atmosphere" at home! Mix different liquids like water, oil, and food coloring in a jar. Watch how they interact, just like the weather on other worlds.

5. Space Snack journey: Plan a space-themed snack journey! Make snacks that look like planets or stars. Use fruits, cookies, and candies to build your very own delicious world.

6. Telescope Time-lapse: Imagine you're a space shooter! Take shots of the night sky from your garden over several nights. Create a time-lapse to see how the Moon and stars move in the sky.

These tasks are your ticket to becoming a space explorer just like the JWST team. Have fun, and remember, the world is full of wonders waiting for interested minds like yours!

Conclusion

Thank you, James Webb, for serving as our space hero! We are very thankful for the wonderful telescope you provide that helps us discover so much about the world.

Dream Big

Now, little space lovers, it's your turn! Dream big dreams about traveling space. Maybe you'll be the one who creates a future telescope or fly to faraway planets. The universe is full of secrets waiting for you to discover. So, let your mind soar and dream those huge space dreams. Who knows what incredible experiences you might have in the future of going into space!

www.ingramcontent.com/pod-product-compliance
Lightning Source LLC
Chambersburg PA
CBHW050759290526
45792CB00008B/2248